# The Manual Screw Press

# The Manual Screw Press

*for small-scale oil extraction*

Kathryn H. Potts and Keith Machell

PRACTICAL ACTION
Publishing

Practical Action Publishing Ltd
The Schumacher Centre, Bourton on Dunsmore, Rugby, Warwickshire, CV23 9QZ, UK
www.practicalactionpublishing.org

First published in 1995
Transferred to digital printing in 2008

A catalogue record for this book is available from the British Library.
A catalogue record for this book has been requested from the Library of Congress.

ISBN 978-1-85339-1989 Paperback

Since 1974, Practical Action Publishing has published and disseminated books and information in support of international development work throughout the world. Practical Action Publishing is a trading name of Practical Action Publishing Ltd (Company Reg. No. 1159018), the wholly owned publishing company of Practical Action. Practical Action Publishing trades only in support of its parent charity objectives and any profits are covenanted back to Practical Action (Charity Reg. No. 247257, Group VAT Registration No. 880 9924 76).

Cover photo from Practical Action Publishing photo library, Rugby
Illustration on page 2 by Debbie Riviere, all others Matthew Whitton
Typeset by My Word!, Rugby
Printed in the United Kingdom

# Contents

# Acknowledgements

The oil processing work in Malawi was funded by the Overseas Development Administration of the UK government and Save the Children Federation (US). The adaptation of the process and equipment involved the community at the Mkhota Rural Centre and the following organizations and their personnel between 1988 and 1990.

SAVE THE CHILDREN FEDERATION (US) MALAWI:
  Jacques Wilmore, Frank and Maria Molotchwa, Charles Sambani

GTZ/NATIONAL RURAL CENTRES PROJECT:
  Dr M. Leopolt, Mr G. Mkhamanga

INTERMEDIATE TECHNOLOGY DEVELOPMENT GROUP:
  Mel Jones, Paul Elkington, Ann Maddison, Anna Sandals (who wrote the first draft of the training manual), Arun Banda, Garry Whitby, Tony Wilce, Ann Waddington

PETROLEUM SERVICES, PO BOX 1900, BLANTYRE, MALAWI:
  Mr A.J. Harken

APTECH, PO BOX AY Amby, HARARE, ZIMBABWE:
  Brian Jones

# 1. Introduction

Vegetable oil is the richest source of food energy, providing twice as much energy as the same quantity of carbohydrate or protein. Groundnut (peanut) oil has an attractive colour, is a high-quality cooking oil with a distinctive flavour, and is especially liked by people who traditionally include groundnuts in their diet. It is therefore a valuable commodity for which there is likely to be a market, particularly in areas where groundnuts are grown, and where supplies of edible oil from national manufacturers are erratic because of climatic or transportation problems.

In some areas, edible oil is still produced using traditional methods such as hand-pounding, either for family use or for sale locally, but in many places both the traditional technologies and local control of the processing and marketing have been lost. The oilseed crops which are cultivated in household gardens or by smallholder farmers throughout the developing world are, in most cases, bought and taken to the cities to be processed using large-scale, high-investment plant in factories which are generally controlled by multinational companies. Centralized production involves high transport and storage costs, and cooking oil is therefore often expensive and it may be unavailable during rainy seasons in rural areas.

Oil processing projects have been set up in Malawi at Mkhota, Mwonsombo, Ekwendeni, Bolero, Lobi, Mzuzu and Lilongwe, and similar technologies are in use in Zambia, Zimbabwe and West Africa. Drawing from this pool of experience, particularly the project in Mkhota, this manual shows how to work out the feasibility of setting up an enterprise, using a screw press for oil extraction. The essential steps involved in the processing of oil are explained.

The supply of raw materials and the market for the end products are the two external factors which are critical to the success of an enterprise. Without an assured supply of oilseed and customers to buy the oil, a new oil-processing business is unlikely to succeed.

## Is there a local demand for edible oil?

When considering setting up a small-scale oil processing enterprise, as with any business, the first step is to check the level of demand for the product. Apart from the obvious long-term financial reasons, this is important because oil is a perishable commodity, and should be stored for no more than six months.

Many farmers can't get oil because it's too expensive or roads are impassable....

oil is only sold in LARGE quantities...

transport is unreliable in wet season so oil can't be delivered to rural areas...

Farmers sell peanuts to big oil factories....

Industrial processing into edible oils and animal feeds...

expensive transport to and from the processing plants...

Figure 1. *Centralized production may deprive farmers of oil from their own produce*

Find out about the market for the oil by:

o observing customers in markets and shops, and talking to people who might be interested in buying your products;
o checking whether potential customers have contract arrangements that would make it difficult for them to change suppliers;
o investigating the other sources of edible oil;
o surveying retail and wholesale prices;
o talking to people who are selling similar products;
o talking to experts (such as extension agents) who may have access to written reports or surveys.

A good way to start is to conduct a survey of consumers, retailers and wholesalers.

Oilcake is a valuable by-product of oil processing and can be used to make sweets, for example, or animal feeds. This can be crucial for profitability and so it is essential to include oilcake in your survey of local demand.

## Raw materials

Approximately 40–50 hectares of groundnuts, at yields of 400–500kg/hectare, will supply one oil press for a year, producing up to 40 litres of oil per day. If these supplies are not assured throughout the year, alternative sources should be investigated before setting up an oil processing business.

Supplies of clean water and fuel will also be required.

In Malawi, pre-shelled groundnuts were supplied as the raw material. It may be cheaper to buy groundnuts in their shells. If it is necessary to shell the groundnuts before processing, the time and labour required for this process must be included in planning. There are many types of decorticating machines available for the different oilseeds, or groundnuts can be shelled by hand.

## Why manual processing?

The softer oilseeds — groundnuts, sunflower, safflower and sesame — and also copra can be processed manually on a small scale at household or village level, which means the start-up and running costs are low compared with larger-scale motorized expelling. Local groups can therefore produce unrefined, pure and palatable cooking oils at a cost which can be borne by rural consumers.

Manual equipment such as the screw press is not suitable for harder seeds such as soya or cotton seed, for which a motorized expeller is necessary.

Where there is a tradition of manual oil processing, the traditional methods such as hand-pounding, water extraction or using a *ghani* (animal-drawn pestle and mortar) are generally time-consuming and arduous. More efficient technologies have been developed for use on a small scale, including the screw press (also known as the spindle or bridge press). The screw press is suitable for operation by women or men, and offers an opportunity for rural people to control the production and marketing of cooking oil.

The process described here is based on the use of the screw press in Malawi to extract oil from groundnuts. With minor adjustments made on a trial-and-error basis, this process is suitable for all soft oilseeds.

## Opportunities and benefits

### Community development

Small-scale oil processing is most appropriate as part of a rural community development programme, and may be subsidized if necessary to enable the project to operate. Such a programme can provide a way for people to:

o work together to add value to local raw materials;
o improve local nutrition, particularly of infants;
o maintain and develop traditional skills;
o build self-confidence;
o develop organizational skills.

Organizing a batch process and marketing a product are life skills that, once learned, can be applied to most productive enterprises.

## Improved nutrition

Oil is a concentrated source of energy which is particularly useful for small infants, as it enables them to get the energy they need from smaller quantities of food than might be possible from a cereal-based diet which is bulky and high in carbohydrates. Groundnut and sunflower seed oils are high in polyunsaturated fats and contain essential fatty acids which are necessary for normal health. (The fatty acid content alters with variety, soil and climatic conditions.)

The value of oil in the diet can be the reason for establishing an oil press for use by local people at a low cost. At Ekwendeni in Malawi the hospital set up a press so that mothers could process groundnuts which they grew themselves to enrich the diet, especially of infants. The equipment is maintained by the hospital and regularly used at a low level. One of the staff is able to train new users.

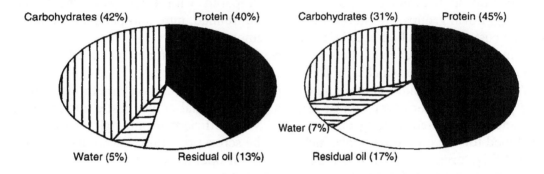

Figure 2. *Composition of groundnut oilcake (range of values)*

Groundnut oilcake, a by-product of manual oil extraction, is nutritious and very palatable. In Malawi, where groundnuts are traditionally used to make a relish called *nsinjilo* (peanut stew), the oilcake can be used instead. It can also be used as a basis for sweets such as peanut brittle and bakery products such as biscuits, cakes and *mandazi* (a type of doughnut made in Africa).

## Employment

Development agencies can encourage groups to set up small-scale oil processing enterprises as job creation projects, for example, by providing the equipment at no cost. The enterprise could employ at least eight people on a full-time basis in various roles, including a sales person, a manager, a mechanic, and manual workers, as well as providing other local employment opportunities.

It would be irresponsible to build up dependence on this enterprise alone, relying on the success of the groundnut crop and the stability of prices, in a community with few resources. However, oil processing on this scale would fit well with seasonal operation in areas where groundnuts are grown, or as part-time employment by women's groups, as occurs at Bolero in Malawi.

## Business creation

A sustainable business enterprise has to be sufficiently profitable to cope with problems of raw material availability, government-controlled price changes, and competitor activity.

If viable, oil processing in rural areas can be a valuable income-generating activity for some of the local smallholder farmers or women's groups and it can provide other farmers with an alternative market. Related business opportunities may emerge, such as animal feed milling, sweet making, bottling and distribution services, which can further strengthen the local community and increase rural control of production.

## Improved supply and lower price of cooking oils

Although large-scale oil production is often subsidized, and oil may be 'dumped' at low prices, an efficiently run small oil processing enterprise should be able to produce oil for local use at a similar price to well-known brands with the advantage of being available throughout the year. Rainy season closure of roads will not affect the supply of this locally produced oil to the rural communities which grow the oilseeds.

# Risks

### Availability of raw materials

The enterprise depends not only on having an adequate market, but also on the availability of the raw materials — in this case, groundnuts — and this is easier to monitor and control if they are grown locally. Operating at the level of 10 batches a day, and with 6 per cent losses, 106kg a day of groundnuts will be needed for one oil press. Therefore, assuming a five-day week and year-round processing, the annual requirement will be over 25 000kg. The efficiency of processing and therefore the number of batches is likely to increase with time, so this requirement must be calculated according to the anticipated throughput. Efforts should be made to have some control over the supply of raw materials, for example by agreeing a contract, so that groundnuts can be purchased monthly.

The quantity and quality of the crop will vary each year, depending on climate, and the pricing of both the oil and oilcake must be reviewed regularly to ensure that changing circumstances and costs are taken into account.

### Aflatoxins

Care must be taken when processing oilseeds because of the danger of aflatoxin poisoning. Aflatoxins are poisonous compounds produced by certain moulds, which may grow on damaged seeds either before or after harvest. These moulds can be prevented by proper drying of the crop and careful subsequent storage, as they cannot grow on material below a 10 per cent moisture content.

Over a prolonged period, exposure to aflatoxins can cause anaemia, liver damage, and even death. This danger means that there has to be a high level of quality control.

If the grading process is followed carefully (see Chapter 3, Step 4) then the danger can be avoided. Humans, animals and, particularly, poultry are susceptible to aflatoxin poisoning.

## Investment and returns

Working out whether a manual oil processing enterprise will be a good investment is the most difficult decision. The payback period, or length of time it takes to cover the initial capital investment, is a good indicator of risk. Most production processes have to be viable at 50–60 per cent capacity to be worth considering.

## Competition

Competitors may take action to keep a new producer out of the market, such as special consumer promotions or price discounts to wholesalers and retailers. In one case, a la. `e competitor purchased all the raw material supplies from the local government marketing centre. There must be sufficient funds to cope with these unforeseen situations. Setting up a new business is always a risk and everyone involved should understand this.

## Government policies

In some countries, smallholder farmers are obliged to sell their groundnut crops to centralized marketing agencies and are thus denied the opportunity to add value to their crop by local processing. This does not, of course, prevent an enterprise from, in turn, buying groundnuts for processing from the central agency, as happened in the project in Mkhota.

Government controls on oilseed and cooking oil prices can make it very difficult for a small-scale business to be profitable. An additional risk is the importation of large quantities of cooking oil at very low prices from a country where production has been subsidized. This practice and the provision of oil as food aid can destroy profitable national oil production.

# The organization of the enterprise

The form of ownership and organization must be considered at the start. The group should include people with appropriate skills to negotiate a loan, purchase equipment and raw materials, and organize production, marketing and record keeping. The amount of time each person can devote to the enterprise should be clear from the beginning, as should ownership and control of the equipment. The enterprise may be co-operatively owned or an individual may be the owner.

A manager to co-ordinate all the activities may be chosen from within the group or be employed by the group. An incentive related to profits can work well as part of the salary. The way the money will be managed — the loan, the income, and the profit or loss — needs to be agreed, along with how this will be checked. Advice on these points may be obtained from local enterprise development agencies.

# What is needed to start?

The first requirements are a lot of determination and a favourable external environment. Given these conditions, small-scale oil processing using a screw press can be a sustainable commercial business.

The main resources are people and equipment (see Chapter 2), raw materials (see Chapter 3, Step 1) and money (see Appendix 2). Assistance may be needed to obtain information about the equipment and about credit agencies which would be willing to lend to this type of enterprise. Some sources of further information are listed at the end of this chapter.

Fuel is required for heating the groundnut flour prior to processing. Wood or charcoal are suitable for this purpose and can be bought by the enterprise.

A secure building which is rodent proof and completely screened is required; an existing building could be adapted, or a new building may have to be constructed. For smallholder farmers or women's groups, a loan may be needed for this level of investment and for working capital. Appendix 2 provides guidelines for preparing the information needed to obtain a loan from a bank or an enterprise development agency.

The required equipment includes the oil press, heating pans, ancillary equipment and a stock of spares, all of which must be costed locally and budgeted for. It may be possible to obtain everything from local manufacturers, which usually reduces the cost of repairs, but if not, an enterprise development agency will be able to help locate a supplier.

The group or individuals wanting to start a small-scale oil processing enterprise should be prepared to make some investment themselves. This may be the building or part of the working capital needed, which may be acceptable in the form of labour and raw materials.

The 'decision tree' (Figure 3) is a summary check-list of some of the important factors to be considered. It is only meant to be a guide: you may be able to think of alternative approaches. The following chapters then set out in detail how to work out the profitability of the enterprise and the process of oil production and marketing.

## Sources of further information

### Bangladesh

ITDG
GPO Box 3881
Dhaka 1000
Tel: +880 2 811 934
Fax: +880 2 813 134

### France

Bureau pour le development de la recherche
sur le oleagineux tropicaux perennes
17 rue de la Tour
75116 Paris

*Continued on page 10*

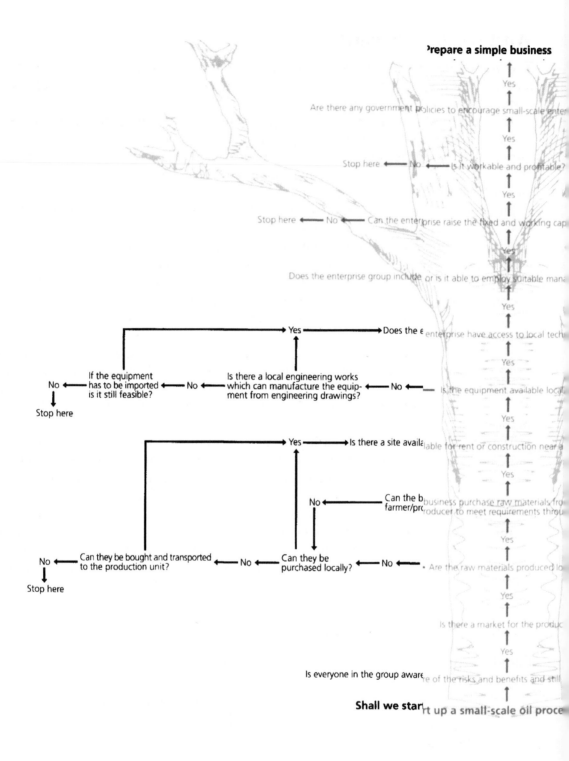

**Prepare a simple business**

↑ Yes

Are there any government policies to encourage small-scale enter

↑ Yes

Stop here ◄── No ◄── Is it workable and profitable?

↑ Yes

Stop here ◄── No ◄── Can the enterprise raise the fixed and working cap

↑ Yes

Does the enterprise group include or is it able to employ suitable mana

↑ Yes

Yes ──► Does the enterprise have access to local tech

↑ Yes

If the equipment | Is there a local engineering works | | Is the equipment available local
No ◄── has to be imported ◄── No ◄── which can manufacture the equip- ◄── No ◄──
is it still feasible? | ment from engineering drawings?

↓ ↑ Yes
Stop here

Yes ──► Is there a site available for rent or construction near a

↑ Yes

Can the business purchase raw materials fro
No ◄── farmer/producer to meet requirements throu

↑ Yes

No ◄── Can they be bought and transported ◄── No ◄── Can they be ◄── No ◄── Are the raw materials produced lo
to the production unit? | purchased locally?

↓ ↑ Yes
Stop here

Is there a market for the produc

↑ Yes

Is everyone in the group aware of the risks and benefits and still

↑

**Shall we star**t up a small-scale oil proce

**Figure 3.** *Factors to be considered before setting up an oil processing business*

8

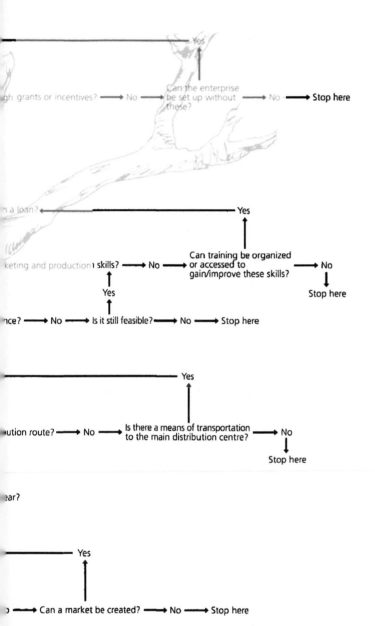

gh grants or incentives? ——▶ No ——▶ Can the enterprise be set up without these? ——▶ No ——▶ **Stop here**

Yes

n a loan? ◀——————————————————— Yes

↑

keting and production skills? ——▶ No ——▶ Can training be organized or accessed to gain/improve these skills? ——▶ No

↑ ↓

Yes Stop here

↑

nce? ——▶ No ——▶ Is it still feasible? ——▶ No ——▶ Stop here

————————————————— Yes

↑

ution route? ——▶ No ——▶ Is there a means of transportation to the main distribution centre? ——▶ No

↓

Stop here

ear?

———————— Yes

↑

⊃ ——▶ Can a market be created? ——▶ No ——▶ Stop here

d enthusiastic to do it? ——▶ No ——▶ Stop here

**erprise?**

9

## Germany

FAKT
Assozierte Ingenieure
Stephan Blattmann Str. 11
78120 Furtwangen

## India

Rajkumar Expeller Corporation
Mr H. Gambhir
Ghat Road
Nagpur 440 018

Tinytech Plants Pvt Ltd
Mr V.K. Desai
Near Bhaktinagar Station
Tagore Road
Rejkot 360 002

## Malawi

Save the Children Federation
PO Box 30374
Lilongwe 3

Small Enterprise Development
Organization of Malawi (SEDOM)
PO Box 525
Blantyre

Malawian Entrepreneurs Development
Institute
Private Bag 2
Mponela

Petroleum Services Ltd
PO Box 1900
Blantyre

## Nepal

Development and Consulting Services
Director
c/o UMN
PO Box 126
Kathmandu

## Pakistan

Dr Akbar Saeed
Head of Applied Chemistry Division
Pakistan Council of Scientific and Industrial
Research Laboratories Lahore
Lahore 54600

## Peru

ITDG
Casilla Postal 18–0620
Lima 18
Tel: +51 14 467324 or 475127
Fax: +51 14 466621

## Sri Lanka

ITDG
33 1/1 Queen's Road
Colombo 3
Tel: +94 1 503786 or 586504
Fax: +94 1 502850

## Sudan

ITDG
PO Box 4172
Khartoum
Tel: +249 11 444 168
Fax: +249 11 452 002

## Tanzania

Iringa Manufacturing Co. Ltd
Mr Augustino Katani
PO Box 931
Iringa

Institute of Production Innovation
PO Box 35075
Dar Es Salaam

Tanzania Engineering and Manufacturing
Design Organization
Mr Msolla
PO Box 6111
Arusha

## UK

ITDG
Myson House
Railway Terrace
Rugby CV21 3HT
Tel: +44 1788 560631
Fax: +44 1788 540270

Natural Resources Institute
Central Avenue
Chatham Maritime
Chatham
Kent ME4 4TB

## USA

UNIFEM
304 East 45th Street
New York, NY 10017

## Zimbabwe

ITDG
PO Box 1744
Harare
Tel: +263 4 796420 or 796409
Fax: +263 4 796409

# 2.  What is needed?

For maximum yields, the groundnuts are first crushed to a coarse flour and then conditioned by the addition of water (approximately 10 per cent by weight) and heated, before being pressed to extract the oil. It is important to have enough space for crushing, heating, and pressing the oilseeds, as well as the equipment, supplies of fuel and clean water, labour, and storage facilities.

## Check-list

- o Building with space outside for heating unit (see below)
- o Decorticator (if needed)
- o Roller mill (or pestle and mortar)
- o Heating unit
- o Oil press
- o Miscellaneous equipment as detailed on the following pages
- o Water supply
- o Fuel supply
- o Labour

## Building

If possible, existing buildings should be used, as this can save considerable expense. However, before a building is accepted, it must be sufficiently large (at least 150 square feet or 20 square metres) and the appropriate steps, described below, should be taken to ensure that it is suitable.

### Security
The building must be secured against theft.

### Site
The area should be checked for potential flooding and rodent infestation.

### Roof
The building should be examined and all entry points for rain or rodents should be sealed. A reasonably sound roof is essential for good storage.

### Floors

A concrete foundation is required for the oil press and a concrete floor is preferable to reduce contamination of oil and oilcake by dust. If there are entry points for rodents, the holes should be cemented over.

### Walls

These should also be rodent proof, with entry points sealed.

### Windows

If windows are to be used to improve ventilation, they should be covered with shutters (at night) and with chicken wire to prevent the entry of birds.

Alternatively, ventilation bricks can be built into the lower part of the walls.

### Hygiene

The building should be cleaned thoroughly before being used as a food store.

## Design of production unit

If a building is to be constructed specifically for the oil production site, it should have all the above specifications. Ideally it should be divided into four sections:

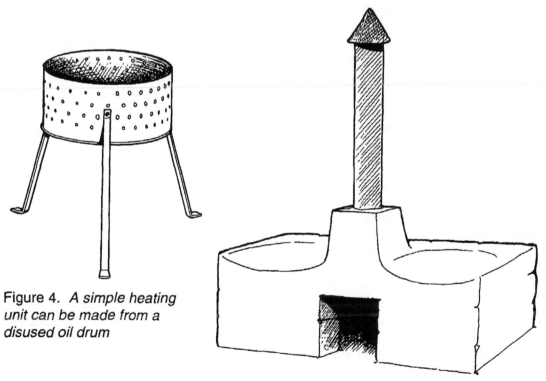

Figure 4. *A simple heating unit can be made from a disused oil drum*

Figure 5. *Brick-built heating unit*

o  Office
o  Production area
o  Storage facilities
o  Shop (if unit is to retail directly to public).

There should also be adequate space outside to construct the heating unit. An oil drum cut in two, with legs welded on, will suffice as a cheap, temporary or mobile heating unit (see Figure 4) but a heating unit made of brick and cement with a chimney is recommended (Figure 5).

The dimensions of the building will vary, but the size of the production unit must be 150–200 square feet to allow space for the extension arms to be inserted in the oil press and turned. The size of the storage space will depend on the quantity of groundnuts expected to be stored at any one time and any other materials which may also be stored there (for example, processing equipment).

A water supply and facilities for washing must also be available.

When planning the production area, give some thought to how materials will be moved around, and lay out the equipment accordingly.

## Equipment

The manual screw press uses some basic extraction technology adapted for local production and maintenance. A similar oil press is manufactured in Gambia.

Engineering drawings are available from IT Publications.

A roller mill may be used to replace the traditional method of crushing with a wooden pestle and mortar. Milling reduces losses and increases productivity which helps to maintain the throughput necessary for continuous use of the press. (See Appendix 3.)

Table 1 lists the equipment you might need, but this is only a guide; a more precise list will be determined by the individuals concerned with the enterprise, on the basis of the available resources and specific needs.

## Labour requirements

At least six people are needed and up to nine could be employed on a full-time basis; more will be needed if some work part time. Labour requirements should be worked out to suit the local situation. Chapter 3 details 18 stages in the production process, for which labour must be allocated.

It is advantageous to work with three teams of two people, for maximum use of facilities. One team is responsible for grading and crushing the groundnuts and two teams alternately press and prepare the crushed groundnuts. If it is more convenient for some people to work part time, they can still work in three teams. However, if the groundnuts must be shelled prior to grading and crushing, a further team will be needed.

Labour for crushing using the roller mill requires two people for two or three hours

# Table 1. Equipment for an oil processing enterprise

| Equipment | Quantity | Use |
|---|---|---|
| Heating unit | 1 | Heating |
| Heating pans | 2 | Heating |
| Wooden stirrers | 4 | Stirring during heating process |
| Oil press | 1 | Pressing of oilseeds and attachments |
| Pestle and mortar<br>or Roller mill<br>with spacers | 5<br>1 | Pounding<br>Crushing |
| 20-litre metal or plastic bucket | 1 | Storage of nut during crushing, carrying water |
| 15-litre plastic bucket | 2 | Weighing of flour |
| 5-litre metal bucket | 2 | Collection of oil |
| 20-litre plastic bucket with sealable lid | 10* | To store small quantities of oil (and selling) |
| 200-litre metal oil drum | 2 | Storage of water |
| 200-litre plastic drum | 1 | To store large quantities of oil |
| 50kg weighing scales<br>(spring balance recommended) | 1 | Weighing of raw materials |
| Bamboo trays | 3 | Winnowing |
| 500g tin cup | 1 | To transfer flour into press |
| Rubber scraper | 2 | To scrape the oil off the tray |
| 5-litre measuring jug | 1 | To measure the oil |
| 1-litre measuring jug | 1 | To measure the water |
| 5-litre plastic bottles | 5 | Filtering of oil |
| Metal funnels varying sizes | 5 | Filtering and selling of oil |
| 20-litre tin bath | 2 | Mixing process |
| Various sized tin measures | 5* | For selling oil |

* Quantity is dependent on choice of marketing strategy


# Table 2. Labour requirements for the main production stages

| Stage of production | Labour required |
|---|---|
| Grading | As many as possible |
| Crushing | 2 mill operators (or 10 pounders) |
| Pressing oil | 4 people (2 teams of 2 people) |
| Selling oil* | 1 person |
| Manager/Supervisor | 1 |

* Other sales people may be required, depending on the choice of marketing strategy.

each day. Once the daily requirement of groundnuts has been crushed, those two people could sort seeds for crushing the next day and help with pressing.

Crushing can be done by traditional methods so long as the labour is available and their wages are costed into your calculations. It is particularly important to ensure that an adequate supply of crushed groundnuts will be available for pressing using this method, as it is slow and laborious.

Labour for using the screw press can be organized into two teams of two people. This allows 10 pressings to be comfortably performed in one day. While one team is pressing the first batch, the second team can begin preparing for the second batch, and so on.

There are different ways of organizing production. An alternative to the system described would be for the same workforce, using a roller mill, to crush a week's requirement of groundnuts on day one, with heating and oil pressing on days two, three, four and five. Up to 15 batches per day have been achieved in Malawi using this arrangement.

## Production management

Apart from the oil pressing, the following stages must be included in processing:

o Storage
o Sorting of nuts/seeds
o Weighing of nuts/seeds for crushing
o Weighing of crushed nuts/seeds and recording losses
o Weighing of flour for processing
o Measuring of water for processing
o 'Handfeel' test during heating
o Measuring of oil after pressing
o Weighing of oilcake after pressing
o Decanting, filtering and heating of oil for storage and sale.

When production is completed each day, the unit should be cleaned thoroughly.

Early experience in Malawi showed that production of eight batches of approximately 3 litres (24 litres a day) could be completed within six hours and, with time, production increased to 12 batches (around 36 litres) in an eight-hour day. It may be possible to produce up to 4 litres per batch, depending on the oilseed variety and with efficiency of production, but when starting up it is better not to overestimate output.

Costings may show that it is necessary to complete more than eight batches to break even. It is up to the group or manager to decide how much oil to produce and how to organize production, but the routine described is known to work well.

As oil processing consists of several stages, it is important to organize production to ensure that each stage can proceed as required without delay.

### Training

Workers should be trained in all stages of the production process. They should be aware of the importance of good hygiene and good storage practices. All should be

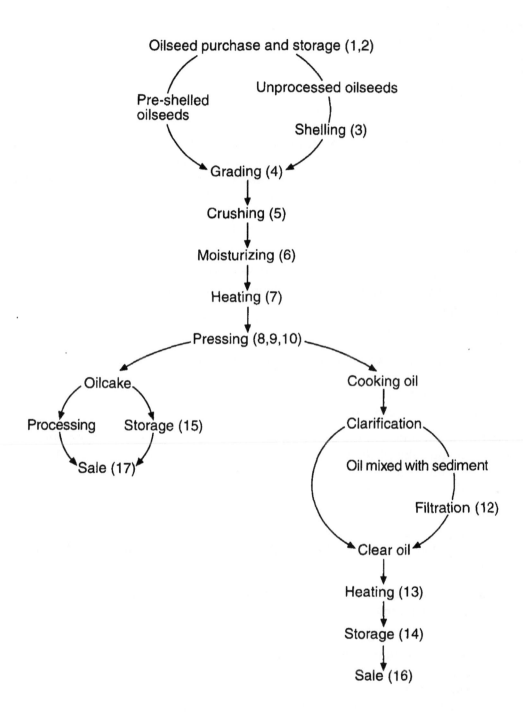

Figure 6. *Flow chart of the production process (each numbered step described in detail in Chapter 3)*

able to perform the 'handfeel' test (see Chapter 3, Step 7) and know how to use the spring balance scales to weigh raw materials.

Both men and women can participate in all aspects of oil processing using the screw press.

## The role of the supervisor

Good supervision can minimize variations in production. In Malawi, the groundnut oil yield varied from 3.0 to 3.7 litres per batch — a range of 700ml. The low yields were attributed to staff apathy and carelessness. A co-operative group may feel more motivated, or workers could be offered incentives to reach targets.

Supervision may be necessary to co-ordinate the workforce and to prevent mistakes, even if the staff are trained in all stages of the production process. During the heating process, when performing the handfeel test, and for all weighing stages of the production process, supervision may be especially important.

The supervisor's duties include:

o Supervision of staff during all stages of production;
o Daily organization of workforce including cleaning duties;
o Keeping records:
  • Daily production schedules
  • Staff attendance
  • Litres of oil and weight of oilcake produced from each batch (daily)
  • Daily sales of oil and cake
  • Stock records of oil, cake and groundnuts (before and after crushing)
  • Cash received and paid.

During the initial stages of the business, the supervisor may also be responsible for the selling of oil and cake (if the latter is being sold from the production site). Once sales increase, an additional person will have to be employed. Below is an outline of the salesperson's responsibilities:

o Promotion of oil and oilcake;
o Keeping records of sales of oil and cake;
o Ensuring that cash received from sales is handed to the manager or banked on a regular basis.

The manager, who might also be the supervisor, is responsible for the overall running of the business, including:

o Staff recruitment and training;
o Purchase of raw materials and equipment;
o Financial management;
o Staff supervision;
o Drawing up and implementing production and marketing plans.

# 3. Oil extraction, step by step

This section describes the whole oil extraction process in detail, presented in a series of steps. Certain procedures are recommended on the basis of production in Malawi and this methodology can be followed by someone with no previous experience. However, as the process becomes more familiar to the operators, it can be adapted to suit local conditions and materials.

If this manual is being used for training purposes, key points in the right-hand column (see over) can be used as a guide and to provide important safety notes. The questions can be used at the end of a training session to review and test understanding.

The hygiene and safety recommendations are important both because oil and oilcake are foodstuffs and because oil may cause slippery floors and burns if spilt. Since both the oil and the oilcake are intended for human consumption, they should be handled in a clean and hygienic manner. Oil is a difficult material to handle, and care should be taken when handling or pouring, to reduce spillage and wastage as much as possible.

## Safety

- o Avoid transferring the oil from one container to another unless really necessary.
- o Any oil which is inadvertently spilt should be wiped up immediately, especially from floors, to reduce the risk of accidents.
- o Take care to avoid burns through bad handling and accidents.
- o To reduce chances of accidents by burning, oil heating must be done at the end of the day, just before everybody goes home, so that when they come the following morning, the oil will already be cool.
- o There should be no running around the oil site at any time.
- o Follow the recommended maintenance procedures for the equipment.
- o Do not leave equipment lying around. Store it properly when not in use.
- o Send samples of oil and cake for regular aflatoxin analysis.
- o Have a small number of chickens fed daily on samples of oilcake as an indicator of aflatoxin incidence, should it arise
- o Animals and poultry should be excluded from the production area.
- o Care should be taken to keep children away from machinery and oil.

# Hygiene

It is important to ensure that all workers pay attention to basic hygiene guidelines:

o No eating in the workplace.
o Washing facilities should be available to the staff, and hands should be washed at the beginning and end of each day, and after visits to the toilet.
o All processing equipment should be cleaned with soap and water at the end of each day and dried, ready for further use. Holes in the oil cage can become clogged with solid material, and this should be removed with a thin metal spike or wire brush daily.
o All floors should be brushed daily and washed once a week with soap and water. Particles of oilcake should be removed and not allowed to accumulate, as they will attract rodents, insects and bird pests. Birds, particularly doves, pigeons and sparrows, are attracted to particles of oilcake and may be a problem. Open areas and windows may require screening with chicken mesh.
o Maintain proper storage procedures.
o When not in use, all equipment should be properly stored. The roller mill and oil press should be covered by sacking when not in use, to protect them from moisture and dust.
o Discarded materials and waste products should be disposed of, or buried, away from the unit, so as not to attract pests.
o If oil is to be sold into the consumers' own containers, make sure that clean containers are used.

# Step 1. Raw material requirements

The best available groundnuts should be used for oil production, to increase yields and reduce the risk of disease. This entails testing for oil and moisture content.

## Oil content

Groundnut crops occur in a number of varieties to suit different soil and climates. The oil content can vary widely from one variety to another, in the range of 28 to

55 per cent for shelled groundnuts. It can also vary within a particular variety depending on the rainfall, soil type and agricultural practices.

There are varieties of groundnuts with a high oil content which are grown specifically for oil production. Where an oil processing enterprise is established, local smallholder farmers could be encouraged to grow these varieties.

What oilseed varieties do local farmers grow at present?

When planning to buy a large stock of groundnuts, it is a good idea to have the oil content measured prior to purchase. Local public health or national standards laboratories can often perform the analysis for a small charge.

Find out where oil content can be measured locally and the cost.

## Moisture content
Groundnut crops for oil processing need to be properly dried. The moisture content should be below 10 per cent and preferably about 5 per cent to minimize the risks of aflatoxin incidence and to ensure good storage properties.

Tests for moisture content (and for aflatoxin) can be made at government health laboratories. Receiving depots of agricultural marketing boards usually have moisture meters.

With experience, it is usually possible to make a close guess of the approximate moisture content, and farmers and rural people are generally good at this. For example, you can get an indication of too high a moisture content in the groundnuts if they stick to the rollers during milling.

If moisture content is too high, the groundnuts should be sun-dried or mixed with drier supplies.

## How to calculate the quantity of groundnuts required
The production method described in this manual is a batch process, with 10kg of decorticated (shelled) raw material processed per batch. The number of batches processed per day will vary, depending upon the availability of raw materials and labour and the local market for oil and oilcake.

Typical oil yield is 3 litres per batch, so you can estimate the number of batches to process each day according to the local demand.

Production levels of 6, 8, 10 and 12 batches per day will require the quantities of shelled groundnuts indicated in Table 3, assuming an average loss of 5 per cent of raw material during grading and 1 per cent during milling. Assuming year-round production, the annual raw material requirements are as shown in Table 4.

**Table 3. Daily groundnut requirements and oil yields**

| Batches/<br>day | Quantity (kg)<br>of graded,<br>milled<br>oilseed | 6%<br>losses<br>(kg) | Total raw<br>material<br>required<br>(kg/day) | Approx. oil<br>yield/day<br>(litres) |
|---|---|---|---|---|
| 6 | 60 | 4 | 64 | 18 |
| 8 | 80 | 5 | 85 | 24 |
| 10 | 100 | 6 | 106 | 30 |
| 12 | 120 | 8 | 128 | 36 |

Work out the daily groundnut requirement for 9 batches per day.

**Table 4. Annual groundnut requirements**

| Batches/<br>day | Total<br>kg/day | Total<br>kg/week<br>(5 days) | Total<br>kg/month<br>(20 days) | Total<br>kg/year<br>(240 days) |
|---|---|---|---|---|
| 6 | 64 | 320 | 1280 | 15 360 |
| 8 | 85 | 425 | 1700 | 20 400 |
| 10 | 106 | 530 | 2120 | 25 440 |
| 12 | 128 | 640 | 2560 | 30 720 |

Work out the annual requirement for 9 batches per day.

## Obtaining a supply of raw materials

It is important to ensure that a sufficient supply of raw materials is available for the planned level of production.

Sources of raw materials will vary from one country to another, but may include:

o Individual farmers
o Agricultural co-operatives
o Private traders or companies
o Agricultural marketing boards
o Ministry of agriculture.

Where can supplies for your enterprise be obtained?

If a supply of groundnuts has to be purchased, buying in bulk quantities can ensure raw material requirements are met. However, it can have a number of disadvantages:

Is transport a problem?

o Increased size of storage facilities require higher capital investment.

o Improved quality control measures are needed to reduce the risk of spoilage during long-term storage.
o Groundnuts should be stored in their shells over longer periods and therefore additional labour is required to shell the groundnuts.
o High fixed and working capital requirements, if the capital is loan financed, may result in negative cash flows and make a small oil processing business unprofitable.

An alternative strategy to ensure a continued raw material supply is to take out a supply contract with a particular supplier, so that raw materials may be purchased at regular intervals, for example monthly. This will reduce capital requirements and improve cash flows and profitability, while helping to ensure raw material supply.

There are benefits to smallholder farmers in producing for local oil processing compared with selling to a central marketing agency. These include cash in hand and reduced transport costs.

# Step 2.  Storage of oilseeds

Good storage practices are very important in reducing losses and ensuring that the quality of the oilseeds does not deteriorate. Deterioration can occur as a result of a number of factors, including:

o Growth of micro-organisms
o Infestation of insects and mites
o Rodents eating the oilseeds
o Human mishandling of oilseeds, causing loss through spillage
o Use of poor storage containers and sacks
o Exposure to extremes of temperature and moisture.

### Storage building

Depending on the facilities available, the storage building may or may not be attached to the production site. However, the building specifications for the storage facilities are the same as for the production unit.

Inside the building, the oilseeds should be stored as follows:

o The seeds should be in sacks which are sealed (tied with string) and free from holes.
o The sacks should be kept off the floor, on a platform of bricks or wooden poles. This is to prevent the possible uptake of water from the ground.

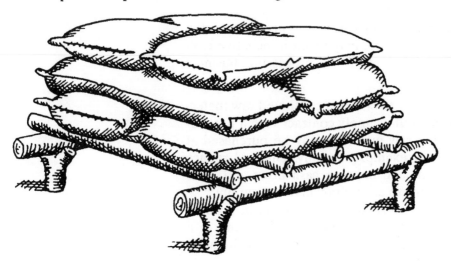

Figure 7. *Oilseeds must be stored in a way that allows air to circulate*

o The sacks should not be stacked against the walls as insects can get into the oilseeds from the walls.
o The sacks should be stacked in a neat manner one on top of another. Space should be left between the stacks so that air can circulate around them.

The oil and oilcake should be stored separately.

## Used storage sacks
Used sacks should be well cleaned and stored separately from new sacks. To clean the sacks:

o Shake the sacks well.
o If the sacks are made of material which can be placed in hot water, boil the sacks or dip them in very hot water, then dry them in the sun. If the sacks cannot be placed in hot water, brush them well and place them in the sun.

## Step 3. Decortication/shelling
In some circumstances (for example, if they are to be stored for long periods) groundnuts may be purchased in

their shells and therefore an initial step of shelling may be necessary in the process. Other oilseeds, such as sunflower, will almost certainly require decortication before pressing. Groundnut shellers are usually evaluated in terms of the percentage of broken kernels produced (in other words, the higher the percentage, the worse the design). Where the shellers are used only in connection with oil processing, this does not matter.

If the oilseeds are to be decorticated on site, the extra equipment and labour requirements must be costed in to the initial calculations.

Figure 8. *Decorticators (manual or powered) can be used to remove shells from oilseeds such as groundnuts and sunflower seeds*

## Step 4. Grading

### Why is grading necessary?

It is necessary to check the groundnuts for damaged, shrivelled, blackened or obviously mouldy seeds, as these may contain poisonous compounds called aflatoxins; which will be carried through the process into the cake and oil.

*Find some aflatoxic nuts and ensure they can be easily recognized.*

## What are aflatoxins?

Aflatoxins are poisonous compounds produced by certain moulds, which may grow on damaged seeds. Over a prolonged period of exposure they can be fatal to humans and animals.

Aflatoxins are not destroyed or removed by heating, or during the subsequent processing stages, and are difficult and expensive to remove if they occur — hence the importance of grading to remove potentially dangerous raw material.

Seeds which are obviously mouldy should be discarded and should not be used. Seeds that have grown mould and have been subsequently dried may not look mouldy in appearance, but may contain aflatoxin. Such seeds are commonly discoloured, shrivelled or obviously damaged and should be removed during grading.

## How to reduce the risks of aflatoxin contamination

Ensure the seeds have been properly dried. Seeds purchased from agricultural marketing boards are often stated or guaranteed to be below a statutory maximum moisture content. However, always grade your raw oilseeds, so that you start the production process with good quality, graded oilseeds.

Teach all workers about aflatoxins and how to recognize bad quality seed.

Store the oilseeds in the correct conditions (see Step 2), primarily in closed sacks off the ground in a cool, well ventilated storeroom.

Ensure good hygienic practices through all stages of production.

Tests for aflatoxins are complex and are generally too expensive to conduct on a daily basis. It is a good quality control measure to have checks for aflatoxins conducted on the oil and cake at monthly or quarterly intervals by a local standards authority, marketing corporation or public health laboratory.

Poultry is particularly susceptible to aflatoxin poisoning and, as any aflatoxins present in the original seed will tend to accumulate in the oilcake, it can be a good idea to feed a small group of chickens daily on a sample of the oilcake as a possible indicator of aflatoxin incidence.

Ensure oilcake is stored under correct conditions. Following processing, mould can grow and infest the oilcake if it is not stored correctly (see Step 15).

The moisture content of oilcake after processing is generally about 4–5 per cent, which is below the level at which the moulds can grow. Should the oilcake become damp, for example during the rainy season, then its moisture content will rise, enabling moulds to grow and aflatoxins to be produced. Oilcake that has become damp or obviously mouldy should be discarded.

If good quality, carefully graded groundnuts are processed into oil in a hygienic way, a high quality product will be produced.

*Describe six ways to reduce the chances of aflatoxin contamination.*

## How to grade groundnuts

Before starting to grade, weigh the oilseeds (groundnuts) using a spring balance scale and write down this weight (A).

*Check that everyone can use weighing scales.*

The oilseeds can most easily be graded by hand. Spread small quantities of seeds on a well-illuminated, light-coloured surface or cloth, and remove damaged or suspect produce by hand and discard it. This can be a communal activity. Stones and plant materials should also be removed at this stage.

The remaining high quality nuts/seeds should be weighed (B) and then stored in closed sacks off the ground in a cool, dry, well-ventilated storeroom. The aflatoxic nuts/seeds should be destroyed immediately by burning.

*If using a spring balance, place an empty bucket on the scales and set the scales to zero.*

Work out the percentage loss at this stage.

## Percentage loss

(A)    10kg nuts/seeds are weighed at start.
(B)    9.5kg nuts/seeds are weighed after grading.
Quantity of nuts/seeds discarded = A–B = 0.5kg

*Grade 10kg groundnuts and work out the percentage loss.*

$$\text{Percentage loss} = \frac{\text{Quantity of oilseeds discarded}}{\text{Total quantity weighed before grading}} \times 100$$

$$= \frac{0.5 \times 100}{10}$$

$$= 5\%$$

Losses are unacceptable above 5 per cent; suppliers should be informed and efforts made to improve quality.

# Step 5. Crushing

It is important that only graded nuts or seeds should be crushed (see Step 4).

## Why crush the nuts or seeds?

o To reduce particle size and increase the surface area to produce more oil.
o To assist in the rupture of oil-bearing cells.
o To increase the compressibility of the oilseeds or nuts within the press, making pressing easier and producing more oil.

Whole nuts or seeds are much more difficult to press and only a small amount of oil will be produced.

There are two ways to crush the nuts or seeds: traditional hand-pounding or roller milling.

Advantages of crushing by traditional hand-pounding:

o A traditional process is likely to be familiar and therefore easier.
o It is labour intensive, thereby creating more opportunities for employment.
o It uses low-cost, locally made equipment which can easily be replaced.
o It reduces the capital investment costs.

Disadvantages:

o The losses are increased.
o It is a laborious and tedious process.
o In order to keep up with production levels of 10–12 batches per day it may be necessary to crush one day in advance of production.
o It produces a groundnut flour of variable consistency which can lead to variations in the final oil yield.

## Equipment needed for hand-pounding oilseeds

Each person pounding will require:

o 1 pestle and mortar
o 1 winnowing basket
o 1 plastic 20 litre bucket.

*How many kilos of nuts or seeds are required for one day's production?*

*Which groundnuts are used for crushing? Graded or ungraded ones?*

*Work out the labour requirement to crush enough seeds or nuts for one day's production.*

Also needed are:

o 3 mats
o 1 set of scales (preferably a spring balance that weighs quantities up to 50kg).

## How to crush by hand-pounding

Hand-pounding uses the traditional wooden mortar and pestle. The graded nuts or seeds are first weighed (make a note of weight A) then pounded into small particles — a coarse flour. This is then winnowed to separate the skins or husks and coarse flour from the fine flour. The coarse flour is pounded some more until the skins or husks can be discarded and a fine flour remains. The flour is then weighed (write down weight B).

In Malawi each of five women was responsible for pounding 17kg of ground-nuts in one day for a production level of 8 batches per day.

## Milling losses

Hand-pounding losses can be high — up to 6 per cent has been recorded. This is caused by:

o Overfilling the mortars
o Careless winnowing
o Consumption of groundnuts by staff
o Inaccurate weighing

One way of reducing the losses is by ensuring good supervision. In Malawi, staff consumption of groundnuts was found to be high, but when this was banned the quality of performance decreased, resulting in more losses.

*How to calculate milling losses*
A = Weight of graded nuts or seeds (kg)
B = Weight of flour (kg)

$$\text{Percentage loss} = \frac{A - B}{A} \times 100$$

Figure 9. *Hand-pounding is slower and more arduous than using a roller mill, but requires less capital investment*

29

## Storage of groundnut flour

Flour can easily pick up moisture from the air and should be stored correctly:

o The flour must be stored in sealed sacks.
o Nylon sacks are preferable to jute, as they are more resistant to infestation.

This applies to both hand-pounding and roller mill methods.

## Stock records

It is very important to know the quantity of nuts/seeds and flour in stock.

List five reasons for keeping stock records.

o If stock levels are known, it is easier to plan future purchases.
o It reduces the chance of running out of stock and so disrupting production.
o If no stock records are kept, missing stock could go unnoticed.
o Stock records aid stock rotation procedures.
o By keeping stock records and calculating milling losses, the losses can be monitored and efforts made to reduce losses.

The use and maintenance of a roller mill for crushing oilseeds is discussed in Appendix 3.

# Step 6.  Addition of water

## Why add water?

o To assist the process of rupturing the oil-bearing cells.
o For more even heating of the groundnut flour.

State two reasons why it is necessary to add water to the flour.

## How much water?

The amount of water added will depend on the oilseeds used. With groundnuts, an addition of 10 per cent water by weight is usual: about one litre of water to 10kg groundnut flour.

Try heating a small batch without water to demonstrate the effect on oil yield.

A little experimentation may be tried to identify the most effective proportion of water for the oilseeds being used.

Figure 10. *The groundnut flour must be moisturized and heated to maximize oil output*

For one pressing of groundnuts, you require 10kg of flour which, for ease of mixing and heating, is divided into two parts of 5kg each and placed in two tin baths.

For each 5kg, add 500ml of water.

o Use a one-litre measuring jug.
o Fill the jug to the top with clean water.
o Tip the water slowly from the jug into the first batch of flour until the water level reaches the 500ml mark when the jug is stood on a flat surface.
o Do not pour the water all in one place, but pour it in the pattern of a snake. (If the water is poured in one place it will be difficult to mix evenly into the flour.)
o Add the remaining water to the second 5kg of flour very slowly.
o Rub the water into the flour with the fingers, ensuring that all the flour is taken from the edges and

What equipment is used when adding water to the flour?

How much water needs to be added to 5kg?

Why is the flour divided into two parts before the water is added?

bottom of the bath. This will take up to five minutes. At the end of mixing, there should be no dry flour left in the bath and it can now be transferred to the heating pans.

# Step 7. Heating

Heating assists in breaking the cell walls and opens the cells that contain the oil. It also decreases the thickness of the oil, allowing it to flow out more easily.

State two reasons why it is necessary to heat the flour.

## How to heat the groundnut flour

At the beginning of the day, the fuel required for the day's production should be prepared. The fires should be well started by the time they are needed for heating the flour. Only a low fire is needed and charcoal can serve well as fuel.

How many people are needed to heat the flour?

The flour should be transferred from the baths to the heating pans, 5kg into each pan. As soon as the heating pans are placed on the heat, stirring must be continuous. This is to prevent burning and to aid uniform heating. It is very easy to overheat the flour, so a close watch should be kept on the fire and it should be reduced when getting too hot by the removal of fuel.

Why is it not recommended to put 10kg of flour into one pan?

How often is the flour stirred?

## For how long should the flour be heated?

The flour's temperature and water content determine when the heating stage is complete and the flour is ready for pressing. The mixture must be heated long enough to reduce its water content and to raise its temperature to about 90°C. This normally takes between 10 and 15 minutes. When ready, the flour should be quite dry and hot to the touch.

How long does it take to heat the flour?

*What if you heat too little?*
There would still be too much water in the flour and this could cause the heated flour to pass out of the holes in the press cage.

*Demonstrate the underheating and overheating of flour to show effects on oil yields.*

*What if you heat too much?*
When black smoke appears, you are heating too much and too quickly. Overheating will produce a dark-coloured oilcake with a strong odour and taste.

*What if you heat with no water?*
If there is insufficient water, charring may occur and the flour will be too dry, so little oil will be expelled.

## The handfeel test
The handfeel test is used to tell when the heating stage is completed. It takes time and practice to master.

*The handfeel test should be practised as much as possible during training.*

When the mixture of flour and water is initially heated, the mixture is sticky and when pressed in the hand or fist will stick together to form a lump or ball of material.

During heating, small handfuls of material should be picked quickly from the heating pan and pressed in the hand.

*State three factors that show when the flour is ready.*

If the material sticks together, it is still too wet and so heating and stirring should be continued.

As heating continues, it can be noticed that the material gradually becomes hotter and drier. Once the material, when squeezed in the hand, does not form a lump and is a free-flowing powder, heating is completed and the material is ready for pressing.

# Step 8. Filling the press

## Why should the press be filled quickly?
The longer the time taken to fill the press, the greater the loss of temperature. Loss in temperature will cause a decrease in the oil yield. In order to prevent this, good production planning is essential. If the start of each batch is properly organized, there should be no delay in the filling of the press.

Figure 11. *Filling the oil press should be done quickly*

Before each pressing, the press cage and press-plates should be placed in the sun to warm. This increases the yield of oil. If the cage is cold, it will absorb the heat from the groundnuts.

## How do you fill the press?

The pans should be removed from the heat one at a time, and the groundnut flour should be transferred into the press using a tin cup.

First, place the cage on the collecting tray, with a press-plate inside the cage at the bottom. This is to prevent the flour sticking to the collecting tray.

Fill the cage with five cups of heated flour and add another press-plate. Press down on the plate and add a further five cups of heated flour. Continue in this manner

How many kilos of flour are placed in the press cage?

Prepare a daily schedule for processing.

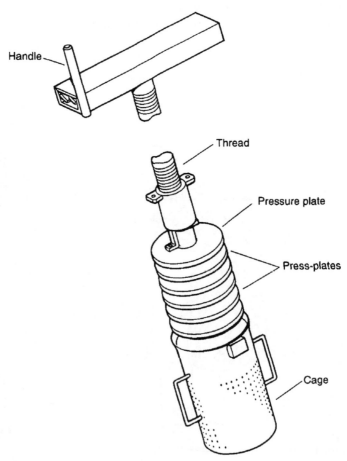

Handle

Thread

Pressure plate

Press-plates

Cage

Figure 12. *Press-plates assist the flow of oil and make it easier to remove the oilcake*

until all the heated flour has been placed in the press cage.

Once all the flour has been poured into the press, place the final press-plate on the top. This is to prevent cake sticking to the underneath of the pressure plate when it is lowered into the press. In total, five divider plates are used.

Wipe any spilt flour off the collecting tray, position the cage carefully under the pressure plate, and place one 3-litre tin bucket under each hole in the collecting tray.

Pressing can now begin.

### Why are press-plates used?

The use of press-plates helps to equalize the pressures within the press cage and assists in the oil flow.

The plates make it easier to empty the oilcake from the cage by dividing it in to several pieces. With the press design and operation described in this manual, it would be virtually impossible to remove the whole block in one piece.

## Step 9.  Pressing

Position the filled cage under the pressure plate. One person should slowly lower the pressure plate by turning the small handle with one hand in a clockwise direction.

Once the handle becomes too stiff for one person to turn, attach the two arms. One person should stand at each end of the attached arms and slowly walk around the press pushing the arms, continuing in a clockwise direction.

The pressing should not be performed too quickly. As soon as oil begins to flow out of the holes of the cage, stop pressing until the flow of oil ceases. Then continue.

Continue pressing in this manner until the pressing becomes too difficult for two people alone.

For the final stages of the pressing cycle, four people will be required.

When practically no more oil flows out, stop. This usually occurs when the fifth thread is reached (that is, when five threads can still be seen above the frame).

To help collect as much oil as possible, rubber scrapers can be used to remove the residual oil from the oil tray.

*Never have more than four people operating the press at one time.*

*Care must be taken not to pick up any grease (from the thread) on the brush. If grease is seen on the brush, this must not be mixed in with the other oil; it must be discarded.*

In addition there can sometimes be some oil above the top press-plate. This can be collected using a clean sponge or paint brush, and squeezed on to the oil tray, or the oil can be poured from the cage.

The oil released from the cage flows on to the collecting tray, which is angled so that the oil can leave the tray via the two holes and can be collected in buckets placed beneath the holes. After pressing, the oil in the buckets is measured in the measuring cylinder and the amount recorded. Yield of oil per batch will of course vary with the quality and original oil content of the seeds pressed, but can be expected to range between 3 and 4 litres per batch.

How long does it take to press one batch?

The insertion of filters will remove any solid material from the oil.

## Problems in pressing

### 'Worms'

During pressing, the heated groundnut flour can sometimes be seen to extrude through the holes in the cage, looking like worms. This can be due to a number of reasons:

o Too much water has been added.
o The flour has been heated for too short a time, so that there is still too much water remaining in the mixture.

### Labour requirements

When the pressing is being carried out, people are often tempted to join in and help out. However, at no time should more than four people be pressing, as this can damage the press.

### Oil squirting from cage

A practical problem sometimes experienced during pressing is a sudden high-pressure squirting of oil from the cage, so that the oil falls beyond the press and is not collected on the tray.

This may occur if the pressure is applied too quickly, or if pressing is continued while the oil is flowing. It

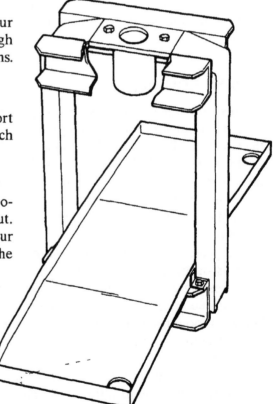

Figure 13. *The oil flows into the angled collecting tray*

also appears to be a characteristic of some varieties of groundnuts.

If the problem persists, even when the pressure is applied slowly, the simple solution is to obtain a thin sheet of tinplate and simply wrap this about the cage, after it is filled, but before pressing commences. Squirting oil hits the sheet and can be collected on the tray.

## Step 10.  Removal of oilcake

### How to empty the press

First, remove the arms. One person should then raise the screw by turning the handle in an anti-clockwise direction.

*Do not rewind the screw too far. This can damage the press.*

Stop when the end of the screw/pressure plate is close to the frame. Lift the cage and hang it on the two brackets on the inside of the frame.

Relower the screw, turning the handle in a clockwise direction. It will probably be necessary to attach one of the arms. When the handle is turned, the cake is pushed from the cage and falls on to the collecting tray.

*Make sure that the cage is attached to both brackets. If the cage is held by only one bracket, it may twist and bind on the screw thread causing it to bend.*

The discs of cake will be stuck to the press plates and can be separated by a metal scraper. They should be allowed to cool before storage. Particles of cake sticking to the plates can be scraped off, cooled and stored separately.

As one person is turning the handle, another person should be collecting the cake as it falls out the bottom of the cage, piece by piece. The cake should be put into a container, not on to the floor.

*How many people are needed to empty the press?*

## Step 11.  Measuring the oil

### Why measure the oil?

It is very important to measure the oil for the following reasons:

o To be able to keep records of the amount of oil produced on a regular basis. These records are essential for running a business. Records of the daily production and subsequently weekly/monthly production figures should be kept and regularly reviewed.

o To assess the performance and extraction efficiency of the press.

o To be able to keep an account of the quantity of oil in stock.

o To aid in stock control procedures.

### How to measure the oil

Collect all equipment together: 5-litre measuring jug, sieve (or tea strainer), 15-litre bucket (with lid).

Place the sieve above the measuring jug. Pour the oil from the buckets that were placed under the press collection trays slowly through the sieve into the measuring jug.

Pour the oil until it is level with the jug's 5-litre mark. Transfer this oil to an empty, clean, plastic 15-litre bucket.

Continue measuring the oil 5 litres at a time until all the oil (from both buckets) has been measured and transferred to the plastic bucket. Note the total quantity.

The oil can then be transferred to another bucket containing oil which has reached the same stage in production.

Place the lid on the bucket, making sure it is labelled *unheated*. It should also be labelled with the date of production.

Place in the storeroom. All unheated oil should be placed together.

*If no measuring jug is available, what could be used instead to measure the oil?*

# Step 12.  Filtering the oil

The newly pressed oil will be slightly cloudy because of the effect of air, and small particles of solid material able to pass through the strainer.

The oil from a day's production should be collected together and allowed to stand undisturbed for 48–72 hours in order to clarify. Most of the clear oil can be carefully poured off or siphoned from the container, and stored ready for heating.

The remaining mixture of solid particles and oil in the bottom of the bucket can then be filtered through a brown paper funnel over several days to remove the final solids from the oil. The solid material after filtering can be added to the next pressing.

Newspaper should not be used as a filter paper, as some printing inks are poisonous and can leak into the

NOTE:
*If groundnut oil is heated after production, without being left to stand undisturbed for 48–72 hours, the oil will be of a dull colour and will froth (foam) when cooking.*

oil. In order that your product has the best possible appearance, care should be taken to ensure that the oil is properly filtered and heated.

## Step 13.  Heating the oil

After filtering, it is recommended that the oil is lightly heated. Heating can improve the oil in a number of ways:

o It removes any residual water from the oil, and thus improves the keeping quality or shelf life of the oil.
o It improves the odour and taste of the oil, as some flavour compounds are driven off during heating.
o It reduces the incidence of any spoilage organisms in the oil.

How do you tell the difference between heated and unheated oil?

### How and when to heat the oil

The filtered oil can best be heated at the end of each week's production.

Place oil in each heating pan and heat gently over a low fire. Make sure that there is someone watching the oil, as it will darken if heated too strongly or for too long.

Once heated, allow the oil to cool in the heating pans before transferring the oil to storage containers, labelled 'heated oil' and dated. Do not let too large a stock of filtered oil build up before heating. Make sure heated and unheated oils are not mixed together.

There are two ways of knowing that oil is ready to be taken off the fire when heating it:

o Using a probe thermometer. Oil is ready at temperatures between 90°C and 110°C.
o Observation. When you see bubbles moving around in the oil, it is ready to be taken off.

IMPORTANT
Oil can become extremely hot and if spilt onto the skin can cause serious burns. Take great care, especially when removing the heating pans from the fire. Do not carry them any distance until the oil has cooled.

Design some posters to illustrate these points and display them in the production unit.

### Safety

Oil is a difficult material to handle, and particular care should be taken with hot oil, when handling or pouring, to reduce as much as possible the danger of burning. Avoid needlessly transferring the oil from one container to another. Any oil that is inadvertently spilt should be wiped up immediately, especially on floors, to reduce accidents.

Remember the following guidelines for safe handling of oil:

o Oil should be heated at the end of the day just before everybody goes home.
o Animals should be kept out of the production area and care should be taken with children.
o There should be no running around the oil site at any time.
o Do not leave any items of equipment lying around.
o Wipe up any spills of oil immediately.
o Follow the recommended maintenance procedures for the equipment.

# Step 14. Storage of the oil

Good storage of oil is very important if the quality of the oil is to be maintained until sale. Good quality oil, if stored incorrectly, can go bad or rancid with an 'off' odour and taste.

Rancidity is caused by chemical reactions between the oil and any water or air present. Therefore, in order to reduce rancidity, contact with water and air should be avoided during storage. As the reactions are speeded up in sunlight, the containers of oil should be impervious to light, or stored away from direct sunlight.

Correctly stored, the oil should remain in good condition for at least six months.

o Any residual water should be removed during heating of the oil, prior to storage.
o Make sure that containers used to store the oil are clean and properly dry, before filling with oil.
o Never top up a container of old oil with new oil. The old oil may be rancid. Always use clean containers.
o Make sure that the container is filled to the top. This is to reduce the air present in the bottle or container.
o Make sure the container has a secure lid or top, to reduce the possibility of spillage or of contamination by water, dust or insects.

## Storage containers

Any suitable sized and locally available plastic container will suffice, from a 20-litre plastic jerry can to a 200-litre plastic drum, provided that:

- o It has a tight-fitting lid or seal.
- o It is clean and free from odours.
- o It is free from cracks.

Many such drums originally carried chemicals, and care must be taken to wash them thoroughly and remove any possible contamination or odour from the containers before they are used for oil storage.

## Storing the containers

The containers should always be stored:

- o Away from sunlight unless they are impervious to light.
- o In a cool, dry place.
- o In a safe place, where they cannot be knocked over.

# Step 15. Storage of oilcake

## Weighing the oilcake

Just as the volume of oil is measured, it is also important to weigh the oilcake produced for the following reasons:

- o To check on the efficiency of the process.
- o For stock control purposes.
- o To be able to monitor losses.
- o For sales records.

The weight of oilcake should include both the weight of cakes or discs produced and the loose cake material or crumbs scraped from the press-plates, as this can still be sold.

## Storage of oilcake

Oil storage guidelines also apply to the storage of oil-cake. The following practices should be applied in the storage of oilcake:

- o Allow the oilcake to cool before bagging and entering into stock.
- o The oilcake is best stored in sacks tied with string.

- o The sacks of oilcake should be stored in a cool, dry place off the ground.
- o Sacks should be labelled with the weight and the date produced.
- o The stock of oilcake should be rotated, selling the oldest stock first.
- o Oilcake keeps well in good storage conditions. Oilcake that has become wet or mouldy should be discarded due to possible risk of aflatoxin.
- o Oilcake is dry and contains some residual oil, so it can easily catch fire. Smoking should not be allowed in the oilcake store or in the processing area.
- o Rats and other vermin eat oilcake, so precautions should be taken against rodent attack and infestation. The area should also be protected against birds.

## Step 16. Selling the oil

There are several options for selling the oil and the choice will depend on the local situation and resources.

### Bulk sales

*Advantages*

- o Reduced cost of packaging material.
- o Lower labour cost for packaging and selling.
- o Simple distribution process.

*Disadvantages*

- o Lower selling price.

### Wholesale bottled sales

*Advantages*

- o Higher selling price.
- o Extra local labour can be employed.
- o Simple distribution process.
- o A brand name can be established.

*Disadvantages*

- o Erratic availability of packaging materials.
- o Additional expenditure on bottling equipment.

- o Higher labour costs.
- o A high standard of quality control has to be maintained.
- o Responsibility for labelling.

## Retailing directly from the processing site

*Advantages*

- o Control over retail price of oil.
- o Option of selling at a lower price if customers provide their own containers.

*Disadvantages*

- o No incentive for local retailers to market oil.
- o The existing marketing chains are not utilized.
- o Expenditure needed for sales staff.
- o Size of local market may be insufficient for all the oil produced.

In Mkhota, the oil was retailed to local customers who brought their own containers to the oil processing site. The oil was sold in 25, 50, 100, 200, 500 or 1000ml quantities so that people with a low income could purchase small quantities as required.

On the basis of this experience, we would recommend that the enterprise sells the oil wholesale using local existing marketing chains. In some countries, such as Malawi, the local retailers will buy in bulk and sell the oil to customers who bring their own containers. In Zimbabwe, this would not be acceptable and the oil would have to be bottled before retailing. Another local entrepreneur or group might be interested in buying the oil in bulk and bottling it before selling on to retailers.

*It is essential to establish the method of selling the oil at the start.*

Remember that if oil is to be sold into the consumers' own containers, you must make sure that a clean bottle or container is used.

Records should be kept of all sales: date, quantity, receipts and customer details.

Regardless of whether the oil is sold in bulk or bottled, before setting up production it is essential to identify:

- o Who will buy the oil.
- o Selling price.

o How you will transport the oil.
o How to promote the oil.

The customers are likely to be shopkeepers and they must be visited in advance of setting up the business to assess their interest and the likely quantity and frequency of purchase. If the business is producing about 700 litres of oil per month and the average consumption of oil is one litre per household/month, then several retailers will serve the 700 households and many will buy other brands of oil. If you assume that 10 per cent of households will buy your locally produced oil, then you will have to supply retailers serving at least 7000 households — so this may mean you will have to travel to many surrounding villages to sell the oil.

## The selling price of the oil
The selling price must be made up of the following:

o Production cost.
o Packaging of the oil (bottle/labour/label/box).
o Transportation to wholesaler/retailer.
o Promotion cost.
o Profit.

When the selling price has been calculated, add on what you estimate the retailers' mark-up will be and compare this price with the products already available locally. If the final retail price of the oil you produce is the same as or higher than that of existing products, then you must offer a product of better quality and presentation backed up with a reliable supply service. If it is much higher it is unlikely that you will be successful, as oil is a basic commodity and price is one of the main factors considered by the consumer.

# Step 17.  Marketing the oilcake

Successful marketing of the oilcake is essential for the enterprise to be successful. The process must be started before production begins.

The most likely market for oilcake is animal feed mills. When oil is produced at large centralized factories, the oilcake by-product is rarely seen on the market. It is sold on directly to large animal feed mills.

In rural areas it may be necessary to find out who buys animal feed and target that market. Another entrepreneur or group may want to set up an animal feed mill, or it may be worthwhile incorporating this into the oil business if no one else is interested in producing animal feedstuffs.

In Malawi, groundnut flour is a traditional item in the human diet and groundnut oilcake can be marketed as a substitute. Where flour is used in a recipe, it is possible to substitute 25–50 per cent of the flour with groundnut cake. This can lower the cost of baked products and increase the nutritional value, as well as improving the taste.

The sale of groundnut oilcake to a local baker or sweet producer would enable the oilcake to be sold in bulk. If the oilcake is crushed before sale, the price should be increased accordingly to cover the labour costs.

Other oilcake, such as that made from sunflower seed, is not suitable for human consumption and its main use will be as an ingredient of animal feeds.

# Step 18. Maintenance of the oil press

The tools needed for maintaining the oil press are:

- Large screwdriver
- 19mm spanner
- 24mm spanner
- Small paintbrush (for cleaning)
- Petrol or paraffin (for cleaning)
- Grease (for lubricating the thread)
- Oil (for lubricating the bearing)

## Oil press maintenance requirements
In order to get the best performance from the oil press and to help avoid unnecessary breakdowns, it is necessary to maintain the equipment as detailed in the following schedule.

*End of each day*

- Wipe oil and groundnut pieces from the collecting tray, press frame and pressure plate.

– Clean the cage and press-plates and store them in a clean, dry place.

*End of each week*

– Wipe away any excess grease from the thread and nut.
– Check that the two lifting ring screws are tight at the top of the bearing assembly.
– Check that the four retaining flange bolts are tight at the top of the bearing assembly.
– Check that the two bolts that secure the thread nut are tight at the top of the frame.
– Remove the four bolts attaching the collecting tray to the frame. Remove the tray, clean away any oil and groundnut pieces and refit the collecting tray.

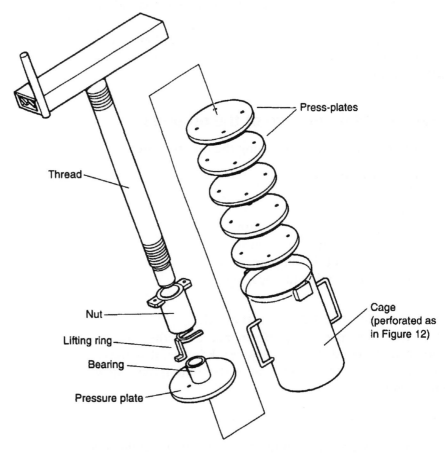

Figure 14. *The oil press*

- Apply fresh grease to the thread. Wind the thread up and down two or three times to distribute the grease evenly along the thread.

*End of each month*

- Remove the pressure plate and bearing assembly from the end of the thread (see procedure A below).
- Remove the thread by winding it to its fully raised position and lifting it out of the frame.
- Clean the thread with a little petrol or paraffin.
- Clean the inside of the nut with petrol or paraffin. Inspect the nut insert for excessive wear. If necessary replace the nut insert (see procedure C).
- Dismantle the bearing (see procedure B below) and empty out its oil.
- Clean all the bearing parts with petrol or paraffin. Inspect the bearing disc for excessive wear.
- Having removed the collecting tray, remove the four foundation bolts and lift the frame off the foundation bolts. Clean away any oil and debris and refit the frame.
- Refit the thread.
- Reassemble the bearing (see procedure B below).
- Half-fill the oil chamber on the side of the bearing with fresh oil.
- Fit the bearing and pressure plate assembly on to the end of the thread (see procedure A below).
- Apply fresh grease to the thread. Wind the thread up and down two or three times to distribute the grease evenly along the thread.

## Dismantling and reassembly procedures
The above maintenance operations require the following dismantling and reassembly procedures:

*Procedure A: Bearing removal and replacement*

Bearing removal:
- Loosen the two lifting ring screws.
- Carefully supporting the pressure plate and bearing assembly, unscrew the lifting ring from the bottom of the thread. Lower the bearing to the ground.

Bearing replacement:
- Raise the pressure plate and bearing assembly to the bottom of the thread and screw the lifting ring on to the thread.
- Make sure that the thread end is touching the bearing disc before screwing the lifting ring to touch the inside of the retaining flange lightly.
- Tighten the two locking screws on the lifting ring.

*Procedure B: Bearing disassembly and reassembly*

Bearing disassembly:
- With the bearing disconnected from the thread end (see procedure A above), remove the four nuts and lift off the retaining flange.
- Remove the lifting ring, bearing and bearing disc from between the four stud bolts.
- Press the bearing disc from the bottom of the bearing.

Bearing reassembly:
- Fit the bearing disc into the bottom of the bearing.
- Place the bearing between the four stud bolts with the bearing disc facing downwards.
- Place the lifting ring onto the bearing (with lip facing downwards) and fit the retaining flange over the lifting ring on to the bearing and stud bolts.
- Tighten the four nuts down on to the flange.

*Procedure C: Nut insert replacement*

The nut insert is designed to wear instead of the more expensive thread. When excessive wear results it will need replacing.
- Remove the two securing bolts and lower the nut from the frame.
- The nut and thread should be taken to a specialist engineering shop that can machine a replacement insert to match the thread and then fit it into the nut body. Alternatively, a new nut insert can be ordered from your local press manufacturer.
- Bolt the nut back into the frame.
- Install the thread and ensure that the thread end

winds down to meet the very centre of the collecting tray before tightening the two bolts.

## General maintenance

All floors should be brushed daily and washed once a week with soap and water. Particles of oilcake should be removed and not allowed to accumulate, as it will attract rats, insect and bird pests. Birds, particularly doves, pigeons and sparrows, are attracted to oilcake and may be a problem. Open areas and windows may require screening with chicken mesh.

Discarded materials and waste products should be disposed of, or buried, away from the unit, so as not to attract pests.

# Appendix 1.  Basic records required

| Type of record/book | Information kept |
| --- | --- |
| 1. **Attendance book** | Names of employees, roll call |
| 2. **Pay roll forms/book** | Names of employees, wage rates, days worked, deductions/additions, net pay, position of employee, signature, date and month of payment. |
| 3. **Daily production book** | |
| a. Crude (unheated) oil | Date of production, names of processors, amount of crude oil produced per batch, number of batches, total crude oil produced per day, day's opening stock of crude oil, crude removed for heating, day's closing stock of crude oil. |
| b. Oil and cake | Day's opening stock of consumable oilcake produced on daily basis, amount of oilcake sold or taken out and closing stock of these products. |
| 4. **Stock control book** | |
| a. Raw materials | Date, opening stock of raw materials — e.g. groundnuts, quantity of groundnuts purchased, amount of groundnuts taken out for flour production, closing stock of groundnuts. |
| b. Crushed oilseeds | Opening stock of crushed oilseeds — e.g. groundnut flour, amount of flour produced, amount of flour issued for oil processing, closing stock of flour. |
| 5. **Daily sales book** | Sales of oil processing products on daily basis at retail prices. May include date, description of product, unit of measurement, unit price, frequency of sale per unit measurement, total amount of money realized and total quantities of products sold. |
| 6. **Credit sales book** (if applicable) | This book may be useful where oil processing products are sold in bulk at wholesale prices to retailers who may act as agency of the unit. May include information such |

as product description, date of delivery, name of agent and location, delivery note number; quantity of product, unit price, total amount, amount paid and outstanding balance. This book, if well maintained, may indicate at a single glance the total amount of money in debt.

7. **Cash book**    Cash received and paid should be recorded daily.

# Appendix 2.  Financial considerations

Factors to be assessed when considering the viability of a business idea include risks, benefits, raw material supply, labour, training, markets, infrastructure and profitability. It may be helpful to produce a check-list for the important factors that need consideration. This appendix outlines the factors that must be considered when thinking through whether to start a business, before covering each factor in more detail. An example is provided only as a guide, and may be modified to suit local circumstances and conditions.

First of all, the capital costs for setting up the business should be worked out, followed by the operating costs and the monthly income, to establish whether there is a need for a loan or credit, and if so how much and when. Obtaining loans or credit can be a long and difficult process, even when there are banks, agencies and governments encouraging the set-up of small businesses. It may be possible to obtain the assistance of a local organization, local official or personal contact to help facilitate the loan application.

An assessment of profit and loss will be necessary, although of course no forecast guarantees the success of an enterprise in practice. The success of a business depends on local conditions, which can change, and decisions must be made on the basis of available information. Any budget or forecast will have to be adjusted once the business is running and the actual situation is known. However, the assessment should be as realistic as possible, as it will be scrutinized when used as part of a proposal for seeking a loan.

It is essential therefore to have a *profit and loss forecast* before starting a business, as this will provide a guide to the potential profitability of the enterprise. It will also serve as a guide to how the business is likely to cope with changes in local conditions (increase in raw materials costs, for example). The effect of possible developments can be assessed by calculations which are called sensitivity analyses. For example, interest rate changes can be applied to given sales and expenditure forecasts to show the effect these have on profits.

Established competitors may try to prevent a new business from succeeding by various means, including cutting prices. An established business may be able to cope with losses for a sustained period, whereas a new business will find it more difficult. All the people involved with the business must be aware of this type of risk.

# Profit and loss

If, as is likely, the potential owners do not have sufficient funds to purchase equipment and operate the business outright, a loan will need to be negotiated.

There are two elements that need to be considered when assessing the size of loan that a business requires: fixed capital and working capital.

*Fixed capital* includes the actual cost of the equipment, including a supply of spare parts, and other related costs such as customs duty, transportation to site, installation, buildings and/or land.

*Working capital* includes all variable costs of operating the business. Once the business has begun, payments will have to be made for raw materials, fuel, labour, transport, and so on. As some, or all, of the payments may be due before income from sales is realized, it is important that serious consideration be given to working out *cash flow*. A contingency fund (say 10 per cent of working capital) should be built in, to cover unexpected items such as price changes.

From information collected, an estimate of the profit and loss can be made by comparing fixed and variable costs against the expected income from the enterprise.

## Fixed costs

Fixed costs remain the same regardless of whether the equipment is used, and may include:

o Capital repayments and/or interest on loans for the purchase of equipment and/or buildings, and so on;
o rent of equipment and/or premises;
o depreciation of capital equipment;
o salaried staff;
o licence and insurance payments.

These costs remain fixed in the sense that they have to be paid whatever the level of production from the equipment purchased. However, they can change: for example, once a bank loan has been repaid, the total fixed costs will be reduced. Another example might be the rent on the premises being increased, increasing the total fixed cost.

## Variable costs

Prices should be obtained locally in the calculation of these costs, which vary with the use of the equipment and the level of production or sales:

o Fuel;
o Labour: manual and technical;
o Maintenance: rates, as a percentage of capital costs, should be expected to increase gradually from around 3 per cent to 15 per cent over the first ten years;
o Transportation of raw materials and finished product: business or contractors?;
o Raw material: farm gate or factory gate prices? (i.e. collected by the business or delivered by the supplier?);

o Packaging materials: bottles, plastic containers, etc;
o Marketing and sales: researching customer needs/preferences; improving product; market stall/stand; advertising by radio or newspaper; searching for sales
   outlets; signboards;
o Taxes: enquire about tax rates at local/rural councils;
o Wastes: disposal on or off site, will there be transportation costs?; Is there a possible source of income from the waste?

## Income
This is the amount of money brought into the business by the direct sale of the oil and
oilcake to customers. Use market surveys and costings to determine the price of the
oil (see below).

## Assessing the profitability of a business
Having established costs, both fixed and variable, and ascertained the sales figures, the
next step is to present this information in a monthly format for the first year, and then
annually for the equipment's life. The example which follows is annual to simplify presentation.

Any potential owner/s should first prepare a *Profit and loss statement* for the business and also a *Cash flow statement*. Together these documents will be required by a
bank, credit or other lending agency to assess the potential of the business. They are
also important to the owner/s, as they enable the comparison to be made of actual
profit and loss and cash balance against those forecast.

The following notes and calculations have been invented for an imaginary processing business and should be used as a guide for preparing profit and loss and cash flow
statements. It has been assumed that the equipment has a ten-year life expectancy and
that a bank loan was required which is to be paid back with interest at a rate of 1800
per year in three years.

## Profit and loss statement

| | Year | | | | | | | | | |
|---|---|---|---|---|---|---|---|---|---|---|
| | 1 | 2 | 3 | 4 | 5 | 6 | 7 | 8 | 9 | 10 |
| *Annual costs* | | | | | | | | | | |
| Fixed | 2900 | 2900 | 2900 | 1100 | 1100 | 1100 | 1100 | 1100 | 1100 | 1100 |
| Variable | 3900 | 3900 | 4000 | 4000 | 4100 | 4100 | 4250 | 4250 | 4500 | 4500 |
| Total | 6800 | 6800 | 6900 | 5100 | 5200 | 5200 | 5350 | 5350 | 5600 | 5600 |
| *Annual income* | | | | | | | | | | |
| Sales | 8000 | 8000 | 8000 | 8000 | 8000 | 8000 | 8000 | 8000 | 8000 | 8000 |
| *Profit/(loss)* | | | | | | | | | | |
| Before tax | 1200 | 1200 | 1100 | 2900 | 2800 | 2800 | 2650 | 2650 | 2400 | 2400 |
| % of income | 15 | 15 | 14 | 36 | 35 | 35 | 33 | 33 | 30 | 30 |
| % of costs | 18 | 18 | 16 | 57 | 54 | 54 | 50 | 50 | 43 | 43 |

In the first three years the profit represents 14–15 per cent of the revenue and 16–18 per cent of the costs. This is just acceptable for a viable production unit, but as profits increase greatly once the bank loan has been repaid it would be a worthwhile venture.

To arrive at the total profit for the assumed ten-year life span of the equipment, simply add the final year's profit to those of all previous years, as shown in the Accumulated profit table.

## Accumulated profit

| | Year | | | | | | | | | |
|---|---|---|---|---|---|---|---|---|---|---|
| | 1 | 2 | 3 | 4 | 5 | 6 | 7 | 8 | 9 | 10 |
| Profit for year | 1200 | 1200 | 1100 | 2900 | 2800 | 2800 | 2650 | 2650 | 2400 | 2400 |
| Cumulative | 1200 | 2400 | 3500 | 6400 | 9200 | 12000 | 14650 | 17300 | 19700 | 22100 |

## Cash flow statement

| | Year | | | | | | | | | |
|---|---|---|---|---|---|---|---|---|---|---|
| | 1 | 2 | 3 | 4 | 5 | 6 | 7 | 8 | 9 | 10 |
| *Funds in* | | | | | | | | | | |
| b/forward | | 1700 | 3400 | 5000 | 8400 | 11700 | 15000 | 18150 | 21300 | 24200 |
| Bank loan | 5000 | | | | | | | | | |
| Sales | 8000 | 8000 | 8000 | 8000 | 8000 | 8000 | 8000 | 8000 | 8000 | 8000 |
| Total in | 13000 | 9700 | 11400 | 13000 | 16400 | 19700 | 23000 | 26150 | 29300 | 32200 |
| *Funds out* | | | | | | | | | | |
| Equipment | 5000 | | | | | | | | | |
| Fixed costs | 2400 | 2400 | 2400 | 600 | 600 | 600 | 600 | 600 | 600 | 600 |
| Variable costs | 3900 | 3900 | 4000 | 4000 | 4100 | 4100 | 4250 | 4250 | 4500 | 4500 |
| Total out | 11300 | 6300 | 6400 | 4600 | 4700 | 4700 | 4850 | 4850 | 5100 | 5100 |
| *Net funds in/(out)* | 1700 | 3400 | 5000 | 8400 | 11700 | 15000 | 18150 | 21300 | 24200 | 27100 |

The cash flow statement and the profit and loss statement are very similar, but with one major difference: the former allows for depreciation of capital equipment. This is not a cash charge against the business, but is a recoverable element representing a saving to enable the owner/s to purchase new equipment at the end of ten years. The difference between accumulated profit at the end of ten years (22100) and the cash balance (27100) is 5000, which is the original cost of the equipment.

## What level of production has to be reached before the enterprise starts to make a profit?

The level of production at which a business makes neither a profit nor a loss is called the *break-even point*. A simple break-even chart can be compiled from data of costs, income and the level of production (calculated in batches) per day, as shown on the following pages.

Notes
1.  It can be seen that the break-even works out at seven batches per day. If fewer batches are produced, costs will be higher than income and a loss will be made. If more than seven batches are produced then income exceeds costs and so a profit will be made.
2.  If there is a change in variable costs then there will be a corresponding change in the break-even point. The example shows an increase in these costs, resulting in a higher break-even point. Similarly, if there is a decrease in the variable costs then the break-even point will be lower.
3.  The effect of a change in fixed costs, in this case a decrease following repayment of the loan after three years, means that the break-even point is reduced substantially to under five batches per day. Again, if the fixed costs increase, the break-even will also rise.
4.  If income decreases as a result of price fluctuations, then there is a possibility of the break-even point becoming too high to sustain the business. On the other hand, an increase in income will lower the break-even point, improving profitability.

It is also possible to work out the break-even point in monetary terms:

$$\text{Break-even} = \frac{\text{Annual fixed costs}}{1 - \dfrac{\text{Annual variable costs}}{\text{Annual net sales}}}$$

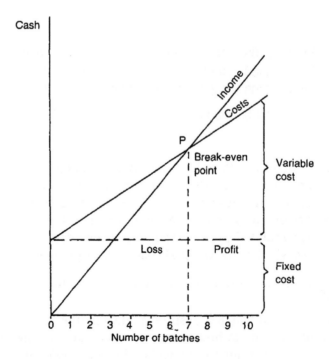

Figure 15. *Break-even chart*

56

| Costs | Year 1 | Year 5 |
|-------|--------|--------|
| Fixed | 2900 | 1100 |
| Variable | 3900 | 4100 |
| Total | 6800 | 5200 |
| Sales achieved | 8000 | 8000 |

$$\text{Therefore, break-even} \qquad \frac{2900}{1 - \dfrac{3900}{8000}} \qquad \frac{1100}{1 - \dfrac{4100}{8000}}$$

$$= \qquad \frac{2900}{1 - 0.4875} \qquad \frac{1100}{1 - 0.5125}$$

$$= \qquad \frac{2900}{0.5125} \qquad \frac{1100}{0.4875}$$

| | | |
|---|---|---|
| Break-even income = | 5658 | 2256 |

Annual sales must be at least 5658 to avoid a loss in the first year and before the business will become profitable. However, in years 4 and 5, after the loan has been repaid, the figure drops to 2656.

If annual sales are less than 5658 in the first year, the business will make a loss. Therefore sales have to be maintained at seven batches each day (163/month or 1952/year).

If a market for the finished product is assured and the production level of ten batches per day is maintained with production losses kept within reasonable limits, then the business will pay the salary of the workers and show a small profit. The benefits to the workers and their families from the income earned and benefits to the community from regular supplies of the product may be considered a sufficient return on the investment.

The most important consideration is whether or not the risk is acceptable for everyone involved. If research before starting the enterprise indicates that raw material supplies can be assured and that markets exist for the finished product this should give a good degree of confidence to take the risk.

## Is profit sufficient?

It might be decided that the profit of 1200 (in the above case) is too little; profit of 1600 or even 2400 may be desired. Of course, these figures must be weighed against the amount the customers may be prepared to pay. Given the same production levels and costs, the owner/s can calculate what the selling price would need to be in order to attain the required profit:

|  | For 1200 profit | For 1600 profit | For 2400 profit |
|---|---|---|---|
| Sales | 8000 | 8400 | 9200 |
| Costs | 6800 | 6800 | 6800 |
| Profit | 1200 | 1600 | 2400 |
| Selling price | 8000 | 8400 | 9200 |
|  | 2760[a] | 2760[a] | 2760[a] |
| = | 2.90 | 3.04 | 3.33 |

*a* is the annual production rate in batches.

The owner/s must use market research to decide whether the product can be sold at the higher batch rates of 3.04 or 3.33 per litre. If consumers will not pay more than 2.90, then increasing income above 1200 will not be possible.

## Is it a good investment?

In practice the income should exceed the break-even point by as large a margin as possible. The 'margin of safety' is an indicator of this and is the ratio by which the budgeted income exceeds the break-even point and the smaller the enterprise the greater the enterprise is at risk.

### Margin of safety

$$= \frac{\text{Budgeted income} - \text{Break-even income}}{\text{Budgeted income}} \times 100 = \frac{8000 - 5658}{8000} \times 100$$

$$= 29\%$$

This is a comfortable margin. Coupled with the fact that production break-even point is seven batches each day (compared with an attainable level of ten each day) this hypothetical case can be considered a viable and sustainable enterprise.

### Three key factors affecting the profitability of an oil processing enterprise

The key factors which affect the profitability of an oil processing enterprise, the raw material supply, the capacity of the equipment and the marketing of the oil and the oilcake, inevitably depend on local conditions.

The main raw material, groundnuts, are usually harvested once a year. If a whole years supply is purchased, then a large amount of capital will be tied up in raw material purchase which will seriously effect cash flow, quite possibly to the extent that the business may become not viable. Another problem here is that the longer the groundnuts are stored, the greater the risk of aflatoxin growth, which must be avoided. Clearly then an accessible supply throughout the year may be prefereable, possibly from a local smallholder or a centralized marketing agency. In many countries there are central grain and seed marketing agencies which purchase all groundnuts produced. If this is the case, it is essential to obtain an agreement to secure regular supplies for small scale oil production before starting the enterprise. In Malawi the central buying agent, Admarc, agreed to reserve a certain quantity of groundnuts for small-scale production.

The capacity of manual screw presses may also be a limiting factor in business. The maximum number of batches that have been processed in a day, using a manual screw press, is 15. It is unlikely that this level of production could be maintained on a daily basis.

Oil and oilcake are both perishable products, and as such should be sold as quickly as possible after production, at most within three months. In some countries, the price of oil is government controlled, in which case the marketing of the oilcake may become critical to the success of the enterprise. It is usually the case that both oil and oilcake have to be sold for a business to become profitable. Identification of customers for both products is therefore of paramount importance.

# Appendix 3. Crushing with the aid of a roller mill

A manually operated roller mill was designed to speed up the crushing process; however, it is still hard work and requires two people for its operation. The roller mill can crush 80kg of nuts or seeds in less than two hours. To ensure that pressing can start immediately in the morning, at least two batches (20kg) of nuts/seeds should have been crushed the previous day.

*Calculate how many hours it will take to crush enough nuts for one day's production.*

Advantages of crushing using a roller mill:

o Less tedious and strenuous than traditional hand-pounding
o Increased productivity to match throughput of oil press
o Reduced milling losses
o Fewer problems with storage of flour
o Consistent flour is produced and so there is less variation in oil yield
o Oilseeds can be crushed on the day of production rather than a day in advance.

*What advantages does the roller mill have compared with hand-pounding?*

Disadvantages:

o Less labour intensive
o More complicated equipment
o Higher capital investment

*What are the disadvantages of the roller mill compared with hand-pounding?*

*Equipment needed*

o 2 clean, dry sacks
o Roller mill
o 2 plastic 20-litre buckets/sacks
o 1 set of scales (preferably a spring balance that weighs up to 50kg).

*How many people are required to operate the roller mill?*

The roller mill consists of two mild steel rollers mounted in a frame. A seed hopper containing the seed to be crushed is mounted above the rollers, and a chute to direct the crushed seed is mounted below the rollers.

One of the metal rollers is fixed, while the position of the second roller can be adjusted by means of a simple screw mechanism, so that the the gap between the rollers can be changed during the crushing process.

The roller gap can be adjusted easily and quickly to the required setting by inserting mild steel plates or spacers made to varying thicknesses — e.g. 2mm, 1mm, and 0.5mm — between the rollers during adjustment.

*Hacksaw blades, 1mm thick, can be used as spacers. Use two blades side by side to set a 2mm gap.*

## Operation

Groundnut crushing is a two-stage operation; first at 2mm and again at 1mm roller gap to produce a fine flour for subsequent processing.

1. Adjust the roller gap setting to 2mm by:

   o slackening off the two adjusting screws fixed to the movable roller
   o loosening bolts holding the bearings and roller shaft to the frame
   o inserting the 2mm spacer between the two rollers, moving the adjustable roller and bearing so that the 2mm spacer is held firmly between the two rollers
   o holding the roller firmly in place, and tightening the roller shaft/bearing bolts to the frame to hold the roller in position
   o withdrawing the spacer.

NOTE: Size and shape of groundnuts differ with variety; gap settings may require some adjustment for the best performance.

2. Fill the seed hopper with graded groundnut kernels, and withdraw the slide plate at the base of the hopper to allow the kernels to fall to the rollers.

3. The first stage of the crushing operation involves the two operators turning the rollers towards each other at approximately the same speed. The partially crushed material is scraped from the rollers by two scraper blades, and falls via the chute to collecting sacks or containers.

How many times do nuts/seeds go through the roller mill?

4. Refill the hopper as necessary until desired quantity is crushed.

5. The partially crushed groundnuts with loosened skins can now be winnowed in order to remove the skins

before further crushing. This is an optional step and has the following advantages:

o Improving the colour, odour and taste of the cooking oil. The darker coloured groundnut varieties can pigment the final colour of the oil and their skins are best removed at this stage. Winnowing may not be necessary with lightly pigmented varieties.
o Improving the taste and appearance of the oilcake. Skins, if not removed, can easily char during the heating stage and produce a burnt tasting, darker coloured cake.
o Marginally improved efficiency and oil yields.

Winnowing can be done by hand using a tray in the traditional manner.

6. Second crushing The roller gap is re-adjusted to a gap of 0.5–1.0mm using the procedure described above but with a 0.5/1.0mm spacer. Crushing proceeds as before.

*Milling losses*
Losses are reduced by using the roller mill, but should be calculated in the same way.

A = Weight of graded nuts or seeds (kg)
B = Weight of flour (kg)

$$\text{Percentage loss} = \frac{A-B}{A} \times 100$$

What are the main points to remember when operating the roller mill?

## Roller mill maintenance
In order to get the best performance from the roller mill and to help avoid unnecessary breakdowns, it is necessary to care for the equipment as detailed in the following schedule.

*Tools*

o Stiff brush (for cleaning)
o Roller groove cleaner (a bicycle spoke is suitable)
o 17mm spanner
o 3mm Allen key
o 0.5, 1.0 and 2.0mm shims (metal spacers) or hacksaw blades, for setting up the roller gap.

*End of each day*
- o Clean roller grooves with the groove cleaner
- o Brush nut/flour deposits from rollers and scraper blades.

*End of each week*
- o Remove feed guide from the feed hopper, and clean nut deposits from the shut-off valve (see B below).
- o Check the Allen screws and roller bearings are tight.
- o Check the scraper blades and adjust them if necessary (see F below).
- o Lightly grease the roller, adjusting screw threads.

*End of each month*
- o Remove feed hopper and guide, clean and replace (see B below).
- o Remove discharge chute, clean and replace (see E below).

## Dismantling and reassembly procedures
The above maintenance operations require various dismantling and reassembly procedures, as follows.

*Dismantling*
A.  Removal of shut-off valve:
- – Remove the rod from its guide.
- – Slide valve out from the feed hopper.

B.  Removal of feed hopper and feed guide assembly:
- – Remove the six bolts at top of hopper.
- – Lift hopper from frame.

C.  Removal of turning handles:
- – Hold roller firmly and turn handle quickly backwards (anti-clockwise) to release it.
- – Hold the handle still, and rotate the roller until the handle comes free from the shaft.
- – Remove the spring washer from the shaft.

D.  Removal of two adjuster plates:
- – Unscrew adjusters until the plates become free.

E.  Removal of discharge chute:
- – Remove the six bolts holding chute to frame.
- – Lower the chute to the floor.

F.  Adjustment of scraper blades:
    – Loosen the scraper bolts.
    – Extend or adjust scraper blades to each roller as required.
    – Re-tighten bolts.
    – Check adjustment of scraper blades to surface of rollers; repeat procedure until properly adjusted.

G.  Dismantling shut-off valve:
    – Remove the four bolts holding valve to frame.
    – Remove the safety guards and feed guide from feed hopper.

H.  Removal of adjustable roller:
    – Remove the four bearing bolts.
    – Carefully lift the roller, shaft and bearings on to the edge of the frame and then down to the floor. Two people are needed for this, as the roller is heavy. Note: Removal of both the adjustable and fixed rollers should be kept to a minimum; for essential purposes like repairs, regrinding of roller grooves.

I.  Removal of fixed roller:
    – As above (H) for adjustable roller.

*Reassembly*
To reassemble, the above procedures should be followed in reverse, but also ensuring:
    – The rollers are aligned with each other; with all eight retaining bolts (four for each roller) loose, the roller gap should be closed completely, and the two rollers aligned before re-tightening the fixed roller retaining bolts.
    – The adjuster plates are correctly positioned; these are left and right-handed and need to be positioned on the correct side for each adjuster.
    – The turning handles are screwed squarely onto their shafts; hold the handle square to the shaft and rotate roller anti-clockwise to engage thread. Tighten the handles by holding the roller and quickly and firmly turn the handle clockwise.
    – Feed guide is correctly positioned. With the roller gap completely closed, position the feed guide up against the adjustable roller before tightening the feed hopper's six bolts.

An environmentally friendly book printed and bound in England by www.printondemand-worldwide.com